I0821669

Sharks

Basking Sharks

by Julie Murray

Dash!
LEVELED READERS
An Imprint of Abdo Zoom • abdobooks.com

1

Level 1 – Beginning
Short and simple sentences with familiar words or patterns for children who are beginning to understand how letters and sounds go together.

Level 2 – Emerging
Longer words and sentences with more complex language patterns for readers who are practicing common words and letter sounds.

Level 3 – Transitional
More developed language and vocabulary for readers who are becoming more independent.

abdobooks.com

Published by Abdo Zoom, a division of ABDO, PO Box 398166, Minneapolis, Minnesota 55439.

Printed in the United States of America, North Mankato, Minnesota.
102019
012020

Photo Credits: Alamy, Minden Pictures, Science Source, Shutterstock
Production Contributors: Kenny Abdo, Jennie Forsberg, Grace Hansen, John Hansen
Design Contributors: Dorothy Toth, Neil Klinepier, Victoria Bates

Library of Congress Control Number: 2019941272

Publisher's Cataloging in Publication Data

Names: Murray, Julie, author.
Title: Basking sharks / by Julie Murray
Description: Minneapolis, Minnesota : Abdo Zoom, 2020 | Series: Sharks | Includes online resources and index.
Identifiers: ISBN 9781532129193 (lib. bdg.) | ISBN 9781098220174 (ebook) | ISBN 9781098220662 (Read-to-Me ebook)
Subjects: LCSH: Basking shark--Juvenile literature. | Sharks--Juvenile literature. | Fish--Juvenile literature. | Ocean animals--Juvenile literature. | Top predators--Juvenile literature. | Carnivores--Juvenile literature.
Classification: DDC 597.33--dc23

Table of Contents

Basking Sharks

Basking sharks are big fish! Only whale sharks are larger.

Basking sharks swim near the ocean's surface. They **bask** in the sun. This is how they got their name.

They are dark in color. Their bellies are paler than their backs.

They have very small eyes.
Their noses are big.

They have large **dorsal fins**. The fin can be 6 feet (1.8 m) tall!

Basking sharks have giant mouths. Their mouths can open 3 feet (.9 m) wide!

Basking sharks are filter feeders. They open their mouths and swim. This is how they catch food.

Water passes through their **gills**.
The food stays in their mouths.
They swallow it in one big gulp!

Basking sharks are gentle giants.

REDBAY STORMFORCE 9.1
9067
CEARBA

More Facts

- Basking sharks are usually 20 to 30 feet (6-9 m) long. Some have grown to be 40 feet (12 m) long!

- They swim slowly. They only move about 3 mph (4.7 kph).

- Adults can weigh up to 11,000 pounds (4,989 kg)!

Glossary

bask – to lie in a warm, calm place.

dorsal fin – an unpaired fin on the back of a fish or whale.

gill – an organ used for breathing by fish and other animals that live in the water.

Index

Online Resources

To learn more about basking sharks, please visit **abdobooklinks.com** or scan this QR code. These links are routinely monitored and updated to provide the most current information available.